David Prince

Plastics and Orthopedics

A Report Republished from the Transactions of the Illinois State Medical...

David Prince

Plastics and Orthopedics
A Report Republished from the Transactions of the Illinois State Medical...

ISBN/EAN: 9783337089368

Printed in Europe, USA, Canada, Australia, Japan

Cover: Foto ©berggeist007 / pixelio.de

More available books at **www.hansebooks.com**

PLASTICS

ORTHOPEDICS:

A REPORT

REPUBLISHED FROM THE TRANSACTIONS
OF THE

ILLINOIS STATE MEDICAL SOCIETY,

FOR 1871.

" It is necessary to the clear elucidation of any special subject, that it
be made so illustrative of the principles which embrace it, as to interest
the intelligent general reader, in its details."—ANON.

BY
DAVID PRINCE, M. D.

DEAF AND DUMB STEAM POWER PRESS.

This Edition of this Report, is presented to his Friends and to eminent Members of the Profession, as a Token of Respect, by the AUTHOR.

~~~~~~~~~~~~~~~~~~~~~~~~~

An edition of this Report, bound in connexion with the former two reports, is published and sold by Messrs Lindsay and Blakiston, Philadelphia.

CONTENTS.

PLASTICS.

ORTHOPEDICS.

LIST OF ILLUSTRATIONS.

INDEX.

REPORT

ON

PLASTIC AND ORTHOPEDIC SURGERY.*

There is a misconception cherished in the medical profession, that the practice of it is like the business of a shoemaker's shop, in which each part of the work is done by a particular operative, the first man selecting the leather and the last one putting on the polish.

The truth is far otherwise. While by *natural selection*, or by accidental circumstances, special attention is given to a particular department, yet the best success in any one, requires a knowledge of all departments, sufficient to see the relations of one to the other. Then again, the general practitioner, outside of large cities, must attend upon most of the surgical cases which present themselves, because a large portion of the persons requiring attention are unable to go to specialists at a distance.

The report on Orthopedics made to the Illinois State Medical Society in 1864, and the Report on Plastics made in 1867, were attempts to make the knowledge of these branches easily accessible, and the present report is a continuation of this effort, with such additions as to include the recent advances. Some considerations and illustrations are embraced in this report which belong equally to other departments of surgery. It was found, however, that they could not be omitted without detriment to the completeness of the comprehension of these two branches.

---

*The Committee on Surgery consists of Dr. E. Andrews, of Chicago, and Dr. D. Prince, of Jacksonville. Plastic and Orthopedic Surgery were assigned by the chairman to Dr. Prince. Other topics of Surgery have been discussed by Dr. Andrews in his current report.

a

PLACISTICS.

Principles.—The constitutional preparation for plastic operations chiefly consists in such treatment as is known to produce a good appetite. This implies a vigorous and healthy process of nutrition, for which the appetite is the mode of asking for the supply. In this condition of rapid nutrition, the plastic exudations are more rapidly solidified with a more vital organization, and the result is a firmer union of parts brought into contact. This general condition is generally best secured by a cathartic which thoroughy clears the alimentary canal a day or two previous to the operation, followed by the use of quinia, or quinia and iron, so as to produce a slight erethism. From two to five grains of quinia may be given in the morning before an operation, and from five to ten, an hour or two before the operation.

It is thought from numerous trials, that quietude and comfort after operations are best secured by anodyne doses which may for an adult be composed of twenty grains of hydrate of chloral combined with a fourth of a grain of morphia. In the practice of the writer, anesthesia is generally secured by a mixture of ether and chloroform, administered upon some very open material, which permits a ready permeation by the air and a rapid evaporation. Nothing answers the purpose better than a patch of a worn out towel of very open texture. It is believed that the tax upon the nervous system occasioned by severe or protracted pain, is greater than that occasioned by anesthesia, and that in addition to the comfort, there is a positive gain for the result.

Local disinfection.—It is exceedingly important to avoid such morbid action in a flap, or along the border of any wound as results in redness and swelling. The use of some anteseptic to prevent or correct the preterefactive change in the exudations is necessary to this end. Carbolic acid is the best of these, and is to be employed so dilate as not to smart.

A solution in water of four grains to the ounce may be employed as a lotion with which the wound can be perpetually moistened.

When, however, it is the purpose to keep the parts dry for

the better retention by means of plasters, the chloride of zinc mixture (mentioned on page viii.) is more appropriate.

One of the most important elements of success in plastic operations is the adoption of such plans, that failure of union will not involve total failure of result.

If the flaps can be taken from such places, and given such adjustments in their new positions as to enable the operator to feel unconcerned, though he sees a total failure of union, he can go quietly forward, adjusting his plasters from day to day, waiting for the union to be accomplished slowly by granulation and cicatrization, which he had hoped to see produced speedily by immediate union.

Size of flap.—There is nothing more necessary than that the flap should be large enough. There is especial danger of deception in those cases in which integument has been put very much upon the stretch by the pull of neighboring cicatrices. An inch of surface will shrivel to a few lines when the distending force is removed; and a flap which had seemed ample, when viewed in its original position previous to its separation, ultimately proves entirely insufficient. The attempt by sutures and pins to prevent this shrinking, generally fails; for either the stitches will cut out, or they will produce such tension of the tissues and compression of the vessels as will result in the gangrene of the flap, through lack of blood to keep up its nutrition.

The base of the flap should be in the direction from which the supply of blood is derived, and the arteries supplying the detached integument should not be divided, if it is possible to preserve them.

The cutting off of the chief channel of supply may cause gangrene of the flap.

Cicatrix.—It is especially important that there should be no cicatricial tissue in the base of a flap, for such is the scarcity and minuteness of its vessels as to make gangrene almost certain. Even when gangrene does not occur, the tendency to shrivel and disappear is exceedingly great.

An illustration of the atrophic influence of cicatricial contraction is given in Ranking's Half-yearly Abstract, Vol. 50, p. 282, for Jan., 1870, from the British Med. Jour. for Dec. 11th, 1869, by T. B. Teale. A symblepharon occured from a burn by a hot cinder, resulting in adhesion of a portion of skin to the globe of the eye. The operation was so made as to leave a portion of skin adherent to the cornea. This gradually became atrophied and transparent, so as not to be perceived at all, except by oblique light. The transparency of the membrane was thus restored.

Inferences are drawn with regard to the uses to be made of this principle of contraction in the treatment of tumors. If the vascular supply can be cut off by causing the vessels to become involved in cicatrices artificially produced, the growth is retarded or arrested. The destruction of a small nevus by inserting the virus of vaccination is upon this principle.

Flap for Transplantation from one part to another.—Is it better that the flap should be attached in its new position immediately; or should it be left for some time, as advised by Taliacotius,* and other old masters, in order that it may be schooled to endurance in the limitation of its vascular supply?

The flap may slough, whether it is stitched to the surface of the face or other part to which it is to adhere, or is left for a while free upon the arm or other part from which it is derived. It is not any more likely to slough when stitched in its new position than when left unattached, provided it is not subjected to any tension. There is, therefore, no time lost from failure of adhesion, in stitching it to its new connection at once. If it sloughs, the case is no worse than if it sloughed without the coaptation and stitching. In either case, a new flap must be taken from a new location.

If the failure is not by sloughing, but by the lack of union by the first intention, the flap can still be kept in position by new sutures if necessary, and union will ultimately occur between the granulating surfaces.

Jasper's case of nose mending, (page xv.), is an instance of union after a strip of cutaneous slough an eighth of an inch in width had occured upon the attached side of the flap. As the slough separated new sutures were taken, and a very narrow subcutaneous adhesion became the base of a cutaneous union, so close, that it was difficult to distinguish the line between the original part and that derived from the arm.

See Plastics, page 34, 1867.

Compression of the arteries of the lips and cheeks.—The great vascularity of these parts renders it important to have the means, (by simple compression), of speedily arresting the flow of blood during and after operations. Figure 1, in the Report for 1867, on Plastics, represents a plan devised for the compression of the coronary arteries. It answers the purpose admirably, and in order to prevent any bleeding on the interior of the cheeks, by a flat compress covering a considerable surface, an instrument has been constructed upon the same principle, and its form is seen here in Fig. 1.

Fig. 1.

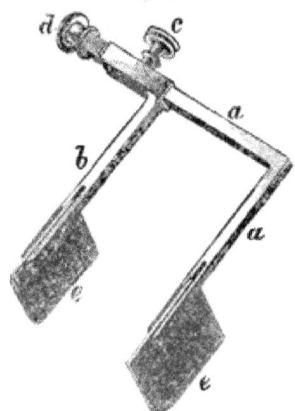

Cheek Compressor. An instrument for compression of the cheek in order to arrest hemorrhage.

a a A shaft bent at a right angle, having at one end a plate *c* and at the other a head *d*, partly for ornament and partly for preventing the escape of *b*.

b A shaft bent at a right angle, the short arm of which is made to enclose the corresponding long arm of the other.

c The mill headed screw; *c* is for the purpose of fixing the two parallel arms *a* and *b* at any distance from each other.

e e Plates of metal for immediate contact in compression. A slot is made in each of the shafts *a* and *b* to receive them, and a rivet is inserted to permit a revolution of half a circle for the greatest possible adaptation. This instrument will also answer the purpose of a dilator in deep dissections as for the femoral or the iliac artery.

The compressor for the coronary arteries is modified by adding the plates at the ends of the compressing arms. These are so made that they will turn through a half circle, thus affording greater adaptability. For steady and reliable compression, a rubber ring is slipped upon the parallel arms where they form an angle with the shaft; and if this ring is too large to produce a sufficient degree of compression, it is folded once or twice or more times, until the requisite force is obtained. It is capable of doing all that can be done by the thumb and finger in compressing a bleeding surface of the cheek, without getting tired, and without being much in the way of the patient's eating and drinking.

Tenaculum Forceps.*—A valuable aid in the operations of plastic surgery is an instrument which will immediately seize a bleeding artery, arrest the flow of blood, and at the same time serve as a handle to the flap until it is convenient, by acupressure or ligature, to permanently close the vessel. The

---

* A description of this instrument with a wood cut was published in the American Journal of the Medical Sciences for January, 1866, p. 147.

tenaculum forceps, represented in Fig. 2, answers this purpose. The instrument is represented as closed. It is first a fenestrated forceps, the upper blade turning upon a hinge which in the figure is concealed by the slide in the central part of the shaft. By the hinge, the upper blade turns nearly over upon the handle spontaneously, as soon as the slide is drawn, being impelled by a spring which in the figure is concealed under the slide. When the upper blade of the forceps is turned over upon the handle, the instrument is a tenaculum, having a point or hook projecting from the extremity of the lower fenestrated stationary blade. The instrument in this condition is applied to a bleeding artery as is done with an ordinary tenaculum. The upper blade is then pushed down upon the lower, and is retained by the slide.

Fig. 2.

a. The handle continuous with the shaft, which terminates in b, the fenestrated lower blade.
c. The tenaculum. point. attached to the lower blade.
d. The fenestrated upper blade.
e. The guard for covering the tenaculum point.
f. The slide which holds the blades in contact.

The instrument then acts the part of a tenaculum and a forceps combined. The blades are made wide enough at the extremity to stop the bleeding from any artery severed in plastic surgery. When the artery is secured by this instrument, the blood is at the same time arrested, so that it is a question of convenience whether or not immediately to employ ligation or acupressure. The compression of a small vessel between the blades of the instrument, for a little time, while the remainder of the operation is progressing, will often permanently arrest the hemorrhage. For the greatest convenience to this end, several instruments of this kind should be at hand. It will be observed, by looking at the figure, that the tenaculum point is protected by a guard which converts the fenestra of the upper blade into two. This is done for the convenience of applying a ligature. When the noose of the ligature is over the largest surface of the blades of the instrument, it is only necessary to draw upon it, when it slides to its position without any possibility of catching upon the tenaculum-point.

To Disinfect, Sutures and Ligatures.—Ever since the establishment of the superiority of silver for sutures, it has been a desideratum, so to medicate the suture of thread, that it may keep its position in surgical favor.

It has also been a great desideratum to obtain the flexibility of linen, silk or cotton, with the unirritating character of silver wire. It is important that the suture should not only lie against the tissues without irritating them, but that it should not have lacerated them in being drawn in. If we can have a suture which is as unirritating to the living tissue, as silver wire, and in addition to this negative quality of not exciting inflammation, is also a disinfectant, (having a positive quality of preventing putrefaction,) we have an improvement upon the silver wire.

The following preparations are proposed as an approach to this end:

Mix together, in the melted state, one part of crystals of carbolic acid and four parts of bees wax.

Mix one part of melted crystals of carbolic acid and eight parts of simple cerate.

Employ strong linen thread thoroughly smeared with the carbolized wax—the smearing of the thread having been done before the time for their use Then, at the time of using them draw the thread through the cerate. This gives them such a facility for gliding, that as sutures, they cause scarcely any friction upon the living tissues, leaving them with the least possible tendency to inflame.

Carbolized sutures, however, can only boast of this superiority for a very short time. In all those cases in which it is desirable that the ligature or suture should remain more than four days, the silver must maintain its supremacy, because in this time the carbolic acid is nearly or quite lost by solution; and as the wax can never completely saturate the thread, the animal fluids absorbed into the suture become putrid and the source of infection. Disinfected sutures which are very superficial, which are intended to secure a very nice coaptation of surfaces—and which can be removed at the end of four days may be employed without any disadvantages and with some increased facility of introduction and removal.

The avoidance of cutaneous ulceration.—In the use of plastic pins and of sutures that are wide-spread and deep, it is found advantageous to protect the cutaneous surface by some means of distributing the pressure, and a disk of lead has been used for the purpose, but some ulceration of the underlying surface has generally been the result.

I have noticed that the absorption of the surface of the skin is prevented or diminished by the presence in the paper compress at the extremities of the pins, of the mixture of carbolic acid, glycerine and chloride of zinc.*

In the operation for entropion, to be noticed further on, in which a silver wire is twisted over a small compress of paper saturated with this antiseptic mixture, there has never been a case of ulceration of subjacent skin.

Galvanic action, in case of the metals, may have something to do in producing the ulceration; and in the case of the paper compress, a part of the explanation may be the prevention of the putrefaction of the fluids unavoidably present under the compress, and the consequent irritation.

Bibulous paper, or several thicknesses of unglazed paper are cut in small squares from one-third to half an inch in diameter, thus: (See Fig. A.) A small hole is previously punched for

Fig. A.

the passage of the needle, unless it is preferred that the exploring needle should make the perforation. One of these paper compresses, saturated with the disinfecting solution is slipped upon an exploring needle, and the needle is made to transfix the parts to be retained in apposition. The plastic pin is then inserted along the groove of the needle; another disk is slipped over the points of the needle and pin; a thread is passed around the heel and point of the pin and needle, as one lies in the groove of the other, and tied across. Another compress prepared in the same way may sometimes be placed over the joining of the wound under the thread, to distribute the pressure of the thread as it passes across the line of incision. After all this, is done the grooved

*Take of a solution of carbolic acid 16 grains to the ounce of water—of a solution of chloride of zinc 24 grains to the ounce of water and of glycerine—each equal parts mix.

needle is withdrawn, leaving the pin in its situation.

Acupressure—The plan of closing arteries and arresting bleeding by acupressure is coming gradually into general favor. The superior advantage of this method over that by ligature, is more in the early period of practicable removal, than in the condition of non-absorption by metalic substances.

It is well known that the irritating influence of the putre-faction within the tissue of ligatures, is scarcely felt until about the time when it is practicable to remove the pins and wires employed in acupressure. From this consideration, the em-ployment of thread or silk, instead of wire in connection with pins, will not be likely to forfeit any of the good results expected from acupressure. The pliability of thread renders it more easy of extraction after the pins have been withdrawn, and the besmearing of the thread first with car bolized wax, and then with carbolized cerate, renders it as harm less as silver wire, as long as the carbolic acid remains un-absorbed. (See page vii.)

If two needles are employed, one on either side of the vessel, the employment of thread in place of wire becomes more practicable. No knot is to be tied around the included tissue, but a bow knot may be made around the heels of the needles which can be untied when they are to be removed.

If, when the needles are withdrawn, (from the second to the fourth day,) the adhesions should have imprisoned the thread, it may be drawn out by pulling at one end, while a wire will not come out until its loop cuts through the tissues.

In addition to the five methods enumerated in the Report of 1867, the following three seem to complete the list of possibly practicable methods of compressing vessels by means of needles. All obscurity will be removed by a glance at the illustrations. These plans are taken from Hutchison's prize essay in the Transactions of the New York State Medical Society for 1869.

To commence the enumeration from the 5th method illus-trated in the Report on Plastics for 1867, p. 12, the first here noticed is the 6th. (See Fig. 3.)

In this plan, the needle is pushed through the tissues by the side of the vessel, and a wire is passed under the point of the needle and twisted over the needle as in the third method.

* Note to page 12 of Plastics, 1867.

This is sometimes more easy of execution and holds the artery
equally well.

Fig. 3.

Sixth method after Pirrie, showing how the wire is put on from Hutchison's **Prize**
Essay in the Transactions of the New York State Medical Society for 1869.

Seventh method (from Hutchison) the second method of our
enumeration reversed.
(See figure 4.) In the
cut the needle is pass-
ing in through the skin
and out through the
same surface, instead
of having its entrance
and exit upon an in-
cised surface. The
object of the proceed-
ing is to hold the
vessel against a bone

Fig. 4.

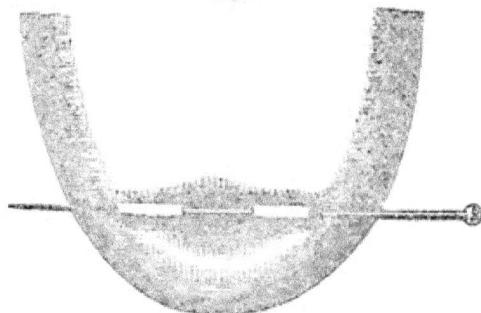

Seventh method. (From Hutchison,) showing the
pin on the cutaneous surface of the flap. The artery
is compressed between the middle portion of the
needle and the bone.

upon which it may lie, i. e., to do what is done by the fingers
or by a compress to arrest bleeding, occurring somewhere in
the distal direction from the place of injury.

The eighth, or Hutchison's method, is a modification
of the third method applied to a wound in which the artery is
not divided, and is applicable in the treatment of aneurism and
in the arrest of hemorrage in branches of a main vessel when
it is difficult to find the bleeding extremities.

The following is Dr. Hutchison's description of the method:

"The artery is first exposed by the usual incision, a loop
of wire about eight inches long is laid in the wound parallel
with and on the side of the vessel next the head of pin (a). (See
fig. 5.) The pin is now carried through the flesh from its cutaneous surface half an inch, more or less, (according to the
depth of the vessel), back from the edge of the incision, so
as to bring it down to the plane of the artery, and then
over the wire and beneath the vessel without disturbing
the vital and organic relation with the nerve, vein or its
sheath.

Fig. 5

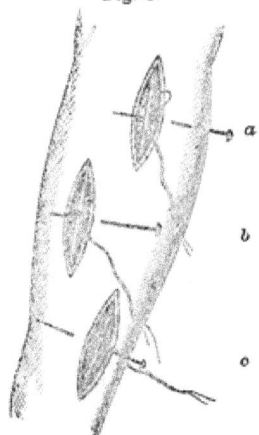

When the point of the pin has emerged from beneath the artery, the wire noose is thrown over the point (b). The point is then pushed through the opposite flap at a point corresponding to that at which it had entered(c). The wire loop is then drawn tightly over the vessel, which is now compressed between the pin below and the wire above. Lastly, the wire is secured by a half turn around the pin, the wound is then closed by metallic sutures."

It will be seen at a glance, that this plan affords all the security of a ligature and admits of early removal. If the carbolized thread is substituted for the wire, the proceeding can be varied in the first part of the process by passing the pin between the threads of the loop, instead of passing it over both as is done with a wire.

The eighth, or Hutchison's, method—substitute for ligature in the continuity of the artery.

Then finally, instead of a twist around the heel of the pin, a knot is tied directly down over the artery. The wound is then brought together and when (after four or five days) the needle is withdrawn, the ligature is readily drawn out as it is superficial to the vessel. It will be found convenient to introduce a grooved needle preparatory to the insertion of the pin, as recommended for sutures.

Epithelial Transplantation.—Much interest has lately been excited, by the demonstration of the possibility of making epithelial cells grow, after removing them from the place of their formation, and depositing them upon another. This is a promise of a great increase of resources in plastic surgery.

M. Guyon* first tried the experiment of planting a portion of detached cuticle upon the surface of a granulating wound. He raised with the point of a lancet, from neighboring sound skin, two minute patches of epidermis, as far as possible avoiding a wound of the true skin. The detachment of the flaps left only a florid spot of skin exposed. He placed them on the center of the wound at a little distance from each other,

---

* American Journal of the Medical Sciences Oct., 1870, p. 555 from Dublin Journal of Medicine and Surgery, August, 1870, from *Gaz. des Hopitaux*, Jan. 11, 1869.

with their deep surface applied to the granulations, and fixed them by two strips of plaster which formed part of the dressing. On the next day they retained their place in spite of profuse suppuration, appearing a little swollen and whiter than when transplanted. On the third day, they were still in place, and M. Guyon raised a larger piece of epedermis, about one tenth of an inch square, and placed it on the wound some distance from them. On the next day, the three flaps were adherent to the surface and could be rubbed without detaching them.

Two days afterwards, the first piece looked pale and thinner and appeared to be extending at the borders. Next day they had united and formed a pale patch, with epedermis forming around them. The third flap was now firmly adherent and surrounded by a little pale band; during the following days, this band, and that around the spot formed by the first two pieces, extended more and more. In a fortnight all these spots were united, and the islet so formed continued to spread rapidly.

A second experiment made on a wound, the border of which had not shown signs of cicatrization failed; the grafts of epidermis falling off after having been some days slightly adherent.

The writer has made numerous experiments in transplanting patches by this method upon healthy granulating surfaces and has failed in all. Dr. Hodgen, as appears farther on, has also failed in attempting to work upon healthy granulations.

Dr. David Page, of Edinburgh,[*] has experimented by shaving off, by means of a bistoury or razor, thin layers of cuticle, superficial to the true skin, and brushing them from the blade of the knife upon the granular surface. A strip of adhesive plaster is laid over the cells thus transplanted.

Dr. Page thinks that "the so-called skin grafting consists not in a transplantation of true skin, but of epithelium only." He says: "It is quite unnecessary to put the patient to the pain of cutting a piece of healthy skin from the body." He thinks there is no increased approach to the character of true skin by this process; that it is only increased rapidity of cicatrization that is gained.

*British Medical Journal Dec. 15, 1870, copied into Phila. Medical news, February 1870.

Dr. John T. Hodgen, of St. Louis, has achieved some gratifying results in the transplantation of epithelium. He says that he has invariably failed in attempting to make epithelium grow upon granulations which are of vigorous growth and bathed with a purulent effusion. It is important therefore, that the ulcer should be allowed to progress until there is very little discharge, and a smooth surface to which to apply the graft.

His experiments have been made in three methods :

1st. The original method of Guyon, that of snipping off the thinnest possible portions of true skin, with the epithelial layer.

2nd. The method of Page, which consists in scraping off the epithelial scales.

3rd. Removing sheets of detached portions of epithelium, and transferring them to the surfaces of ulcers not inclined to heal

This is his mode of proceeding according to the first method : A very fine cambric needle, fixed in a handle, is passed through as small a piece of skin as possible. A sharp knife is passed, with a sawing motion, under the needle, shaving off a portion of skin of the size of a canary seed, transfixed by the needle. The needle with the graft upon it, is then laid upon the ulcer in the same position as upon the skin. The back of the knife, after being dipped in water is then laid upon the needle as it is withdrawn. A cerate dressing is then applied, and the part is not disturbed for several days. At the end of two weeks the graft becomes apparent, and at the end of a month it becomes as large as a finger nail.

The second method consists of scraping the scales of epithelium from another part of the body, and dusting them upon the granulating surface. These are the scales which are usually thought to have lost their vitality and to be only foreign substance to the living tissues. According to these experiments, they seem to be capable of multiplying and forming continuous sheets of new epithelium more rapidly than the large grafts employed, according to the first method.

The third method consists of the application of sheets of epithelium which have become spontaneously detached.

This third method is supposed to be original with Dr.

Hodgen, and only one experiment had been made at the time his report was furnished. It remains to be seen from future observation, how far physiologists have been mistaken in supposing that these dried epithelial scales were dead, like the outer bark of a tree.

The accompanying cut, (Fig. 6), gives the appearance thirty days after grafting, by the first method, upon an ulcer upon the leg of a negro.

Fig. 6.

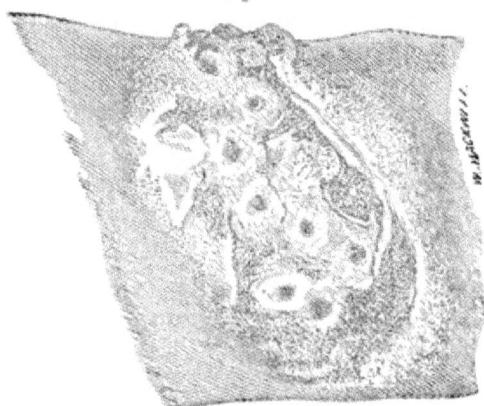

Dr. Hodgen's case of skin grafting engraved from a photograph taken 30 days after implantation.

Dr. Andrews has had a few successes, among many failures, in cuticular transplantations. He has followed only the first method.*

Dr. W. H. H. King has succeeded in transplanting cuticle, by the first method, upon an ulcer upon the back of a negro, who, some years ago, received a burn which is sometimes healed and sometimes an open ulcer. For observation the pigment of the negro is a great advantage.

SPECIALTIES.

Rhinoplasty—a new method by John Wood of King's College Hospital, London.—†From the Half Yearly Abstract, July, 1870, p. 173—from the Lancet for Feburary, 1870.

The following ingenious method is worthy of attention:

A portion of the upper lip is employed in forming the new

*See Report on Surgery in Current Transactions.
†Note to page 40, of Plastics, 1867.

nose, by involuting both its mucous and its cutaneous surfaces.

The upper lip is first cut through vertically by two parallel incisions, extending up from the vermilion border, each a sufficient distance from the median line. The middle portion of the lip having been turned up, the mucous membrane, with half the subjacent substance is dissected off, but is not detached at the vermilion border, in order to make a long raw surface which presents forward when the lip is undoubled and laid over the cavity to be filled. The upper end of this flap is stitched to a surface made raw above the opening to be covered. A large flap is then dissected from each cheek, the base of which is toward the inner canthus, for the supply of blood. These flaps are brought together in the middle line, the raw surface above, coming in contact with the raw surface of the upturnedlip. The remaining portions of the upper lip (after extensive dissection of the cheek from the superior maxillary bone) are brought together as is done in hare-lip. The space left upon the cheeks are covered as much as possible by sliding the integument. Tubes through which respiration can be carried on are placed under the flaps, one on either side of the septum which has been made by turning up the section of the lip. These maintain the contour of the part, until some degree of solidity has been attained. The result is said to have been very satisfactory.

Rhinoplasty—filling a large semi-lunar gap in the left ala of the nose.—H. Jasper, received several years ago, a mutilation of the nose on account of which several plastic operations have been performed by another operator. He desired as little scarring of his face as possible, and after a statement of the different methods, he decided to have a flap taken from his arm.

Dec. 27, 1870. Having taken a cathartic the day preceeding, a flap was dissected from the left arm above the elbow, with the base toward the shoulder, and attached by silver sutures and pins to the defective portion of the nose previously denuded. Strips of isinglass plaster were applied, and these were then and afterward frequently smeared with the antiseptic mixture of chloride of zinc, glycerine and carbolic acid already mentioned. (See page viii.)

The operation was done in sleep from inhalation of a mixture of ether and chloroform. The arm was thereafter maintained against the head by Taliacotius' bandage.

28th. Has taken 100 grains of Hydrate of Chloral producing sleep—ate a few small crackers and drank through a tube, the mouth being inaccessible for cup or spoon.

29th. Took 100 gr., 31st, 35 gr., Jan. 1st 35 gr. of Hydrate of Chloral

Jan. 2d. (6th day.) Removed the plasters and stitches. Found a margin of cutaneous slough embracing the sutures, except one of the two pins which held fast.

There is some deep adhesion. Removed the slough and introduced three silver sutures, bringing the cutaneous granulating edges together.

Jan. 4th. New stitch introduced as a measure of precaution—good appetite.

Jan. 7th. A transverse cut was made upon one edge of the flap, to accustom its circulation to a narrower isthmus.

Jan. 8th. There is union along half the line of opposed incised surfaces. An additional silver stitch was introduced through the flap under which it was retained by a button upon one end, and held at the other end by a plaster over the nose.

Takes 30 grains of Hydrate of Chloral every night.

Jan. 11th. Another stitch.

Jan. 13th. Introduced another stitch, and the cut previously made on the edge of the pedicle of the flap was extended, preparatory to its final separation.

Jan. 14th. (18 day.) Severed the flap from the arm, refreshed the edges which had not adhered, put two stitches through this portion, and applied small strips of plaster as before.

Jan. 15th. The lower edge of the flap appeared gangrenous.

Jan. 18th. The greater part of the flap found to have mortified, and the dead portion was cut away from the living.

Jan. 23d. The small patch of tissue derived from the arm looked well, but the notch was not sufficiently filled up.

The health being good; another operation was performed for the purpose of completing the obliteration of the notch by material derived from the nose itself.

The first variety of the second method,* that is the sliding of a flap in a curved line, was resorted to. A curved incision was made along the natural groove which connects the ala with the body of the nose, permitting the ala to be brought upward and forward to fill the notch. The accompanying outline will afford a conception of the plan. (See Fig. 7.)

This operation failed, in consequence of a slough of the tissue back of *b*, though some gain was made in reducing the depth of the notch.

April 18th. Jasper returned. The cicatrization being complete and the redness of recent inflammation having disappeared, the operation was repeated. The sutures and the pins were removed on the 5th day—a complete success.

Fig. 7.   Fig. 8.

a. The point of the nose. *b.* Notch.
The black line behind *a, b,* shows the contour of the front of the nose.
c. Slit made behind the notch. In Fig. 8, the line *c* becomes straight.
d. An incision along the groove between the ala and the body of the nose in a curved line, commencing at a point bisecting the slit *c.*
e. The flap, the elevation of which is to crowd out the substance behind *b,* and fill the notch. *f.* The point of the flap. *g.* The upper end of the slit *c.*
The elevation of *f* to *g* in figure 8 shows the completion of the operation.

Fig. 9.

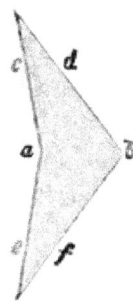

Canthoplasty.[*]—A plan of elongating the palpebral fissure is illustrated by Fig. 9, from an article upon granular lids by Dr. Raschaupt in the American Practitioner (Louisville), for February, 1871. This is the mode of operating. "The lids being kept well apart by means of a spring speculum, the external commissure is divided by a knife or scissors to the extent of three or four lines. The wound presents the form seen in the diagram. The conjunctiva at *a* is then united by a delicate suture to the integument at *b*. In the same way *c* is united to *d* and *e* to *f*."

Canthoplasty from Raschaupt.

---

*Note to page 53 of Plastics 1867.

The practical objection to the plan is, that the mucous membrane; already too small for the space it has to cover, is required to cover still more space. The result will be that the stitches will cut out, or the deformity will be gradually reproduced in consequence of the tension, acting upon the surfaces, as is the case with a cicatrix at the bend of the elbow or between the fingers, or else, the cutaneous surface must be drawn in by the tension of the mucous membrane. The last is an improbable result. As a remedy for granular lids on the principle of relieving the tension of the mucous membrane, this plan of operating is defective in theory and must generally fail in practice.

It was in consequence of repeated failures by this plan of operating, that the idea was forced upon me, of turning in a tract of skin to supply the deficiency of the mucous membrane. The importance of having the inverted skin free from the minute hairs which cover most of the face, makes it necessary to take it from the upper or the lower lid. Fortunately, this necessity coincides with the indication generally present to remove the inversion of the lid. The lower lid is the place of choice, but the flap can without difficulty and without subsequent deformity be taken from the upper lid, or first, a flap can be taken from one, and after adhesion, another flap can be turned in from the other lid, and this can be repeated several times if there should in any case be any necessity for it.

The cuts, (Fig. 10 and 11,) are reprinted from the report of 1867, in order to make this report more nearly complete in itself on this subject.

Fig. 10, represents the external commissure contracted by cicatricial shrinkage following protracted inflammation. The space from which the flap has been taken, is seen below the outer part of the lower lid, and the flap itself is turned up with the wire passing through it, which is to be the suture to hold it in its new position.

A dotted horizontal line, shows the place in which a deep incision is made under the base of the flap, to extend the external canthus. A free dissection, which cannot be shown in the figure, is made in the mucous membrane behind the outer part of the upper lid for the reception of the flap, when turned in to occupy the place of the mucous membrane.

Fig. 10.

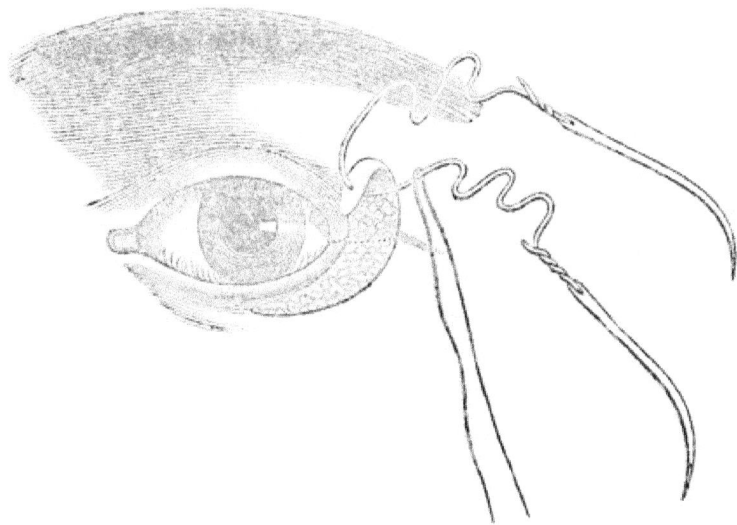

Entropium—the deformity and the new operation for its relief.

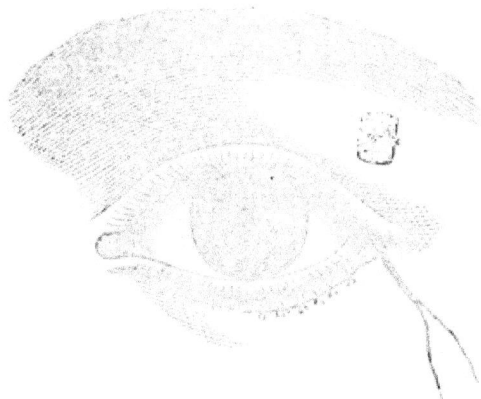

Fig. 11. The operation completed

Fig. 11, represents the operation completed.

The flap has disappeared behind the upper lid, and the retaining suture passes out under the brow and is twisted over a paper compress, disinfected by soaking in the mixture of chloride, of zinc, glycerine and carbolic acid.

The ligature; seen to be twisted and hanging out at the external canthus, is for finally extracting the loop of silver wire, but it is not necessary as the wire, (like a metalic suture introduced any where else,) can be straightened and withdrawn.

It has been learned by experience, however, that this reserve ligature for the withdrawal of the loop of wire, is of great utility in preventing the occurrence of an abscess in the lid. One has never occurred under the observation of the writer

with the presence of this conductor for pus, while the formation of an abscess has not been infrequent, in its absence.

The integument beneath the lower lid, adjoining the space
from which the flap has been dissected, (Fig. 11.) has been
brought together and retained by interrupted sutures.

Spasmodic Trichiasis.—The importance of avoiding or removing the inversion of eyelashes though only spasmodic, is
not sufficiently appreciated as a means of controling acute and
subacute inflammation.

The inflammation precedes; then there comes great intoler-
ance of light. This induces, by reflex action, an extreme
spasmodic contraction of the orbicularis palpebrarum and
in some persons the result is an inversion of the eyelashes, more
often of the lower lid. The gliding of the little hairs upon the
corneal and the conjunctival surfaces aggravates the inflammation, and this in turn increases the spasmodic action.

It would suffice to divide the muscular fibres at the external
angle by a simple incision, (as in Fig. 9.) only that in those
severe inflammations which have already continued several
months, the closure of the incision will be complete and the old
contraction re-induced, before the subsidence of the inflammation. It has been found most satisfactory to perform the
operation by inversion (Fig. 10 and 11,) first recommended to
the profession for the cure of permanent trichiasis.

In the cut, the flap is taken from the lower lid and turned in behind the upper, but as in the operation for permanent trichiasis
the location can be reversed. This not only answers all the
purposes of a simple incision, but it renders the recurrence of the
spasm, by too speedy healing, impossible.

Case.—Mr. J. C., aged about 40 years, has had general inflam-
mination of both eyes for a long time, and lately has had
hypopyum or a purulent collection in the central part of the
cornea of the right eye. There is an inversion of the ciliæ of
the lower lid keeping up inflammation of the cornea. This,
exists upon both sides, but is greatest upon the right.

There is a deep granular condition of the lids, and a general
deep redness of the vascular tissues.

Oct. 29th, 1870. Under anesthesia the muscular fibres relaxed
and the trichiasis entirely disappeared. The operation by inversion was performed, and on recovery from the anesthesia, the

ciliæ stood out naturally and did not again become inverted.

The margin of the cornea was at the same time punctured; the cataract needle passing far enough in, so that by the deflection of the handle, the ciliary band was divided on the withdrawal of the instrument, with the escape of a portion of the aqueous humor.

Nov. 11th. The inflammation is fast disappearing in both eyes, and the angle of the eyelid operated upon, is without deformity. The result has vindicated the propriety of the practice. The cornea was for the time being relieved from the irritation of perpetual scratching by the eyelashes, and in addition to this advantage, the danger of a permanent trichiasis is averted. In the progress of the recovery, the right eye, which had been the worse of the two, gained upon the other.

This patient went home in February, so far recovered as to go about upon his farm, but after a month of improvement, he was obliged to return on account of a relapse, but there has been no return of the trichiasis. May 16th. The patient is in process of recovery. There has been no return of trichiasis. June 16th. Recovery nearly complete.

M. C., a well grown girl 12 years of age, has had inflamed eyes for four years. The sclerotic is deeply injected; the straight or radiating pink lines encircling the iris in a beautiful zone. Two out of these four years, she has been blind or incapable of opening her eyes to the light, and sympathetic sneezing still accompanies every attempt to do so. The examination had to be conducted under anesthesia to suppress the sneezing, and to relax the orbicularis muscle so that the lids could be opened. The ciliæ of the lower lids were considerably inverted, and those of the upper lid slightly so. The operation by inversion, (Fig. 10 and 11), was performed Dec. 19th, 1870. This was done for the double purpose of relieving the cornea of the constant irritation of the contact of the ciliæ and at the same time, of suspending in part the pressure upon the globe, arising from the spasmodic contraction of the orbicularis.

In this case, and also in the case of J. C., there had not come to be a permanent trichiasis, but only one dependent upon continued muscular action.

There was in the present case some shortening of the palpebral fissure previous to the operation.

The case progressed to a complete cure without interruption, and the patient returned home February 9th, 1871.

A. B., a well-grown Miss of 14, has had inflamed eyes for six years; and a part of the time she has been blind from excessive sensitiveness to light. The sclerotic membranes are very vascular and both lids of both eyes deeply granular, without inversion of the eyelashes, but with shortening of the palpebral fissures.

Feb. 28th, 1871. Anesthesia from chloroform.—A seton introduced into the back of the neck, and implantation behind the upper lids of skin taken from below the lower eyelashes (as in Fig. 10.)

This is done upon the theory that the restoration of the healthy vascular condition of the lids will be rendered easier by such implantation of integument behind the upper lid, as will diminish the tension arising from the cicatricial contraction incident to the long continued granular growth.

March 2d. Two grains of Quinia and one of Iron (by hydrogen) every morning.

A powder twice a week composed of Podophyllin $\frac{1}{12}$ gr., Calomel $\frac{1}{2}$ gr., Leptandrin 3 gr., and white sugar 3 gr.

Four drops of Tinct. Verat. virid. at 2 and at 4 p. m. every day.

March 4th. Fourth day after the operation. The fine thread sutures upon the lower lid removed.

March 13th. Removed the silver wire from the lids.

16th. Commenced the daily use of the citrine ointment upon the lids.

18th. Commenced the use, three times a day, of a mixture, so composed, that a dose contains 10 grains of sulphate of magnesia $\frac{1}{4}$ gr., sulphate of iron and 5 drops aromatic sulphuric acid.

April 14th. Went home for a vacation, so far improved as to be able to face the open air daylight without pain.

24th. Returned after 10 days' absence, having still further improved during her home visit.

25th. The operation of inversion repeated upon the left eye in consequence of the want of sufficient opening of the outer canthus, i. e., there was some approach to the condition of the canthus before the first operation.

This was done under chloroform. The general treatment already given, continued.

It will be noticed that this patient took, for a long period, two or three grains of Quinia every morning and eight drops of Tincture of Veratrum viride every afternoon.

It is found by experience that Veratrum does not diminish the appetite when given short of nausea, or even though there may be nausea every day, if the duration of the nausea is confined to short periods. Hence, it is practicable to obtain the general tonic effects of iron and quinia with a diminished frequency of the pulse, and with contracted capillaries consequent upon the use of Veratrum. The combination of sulphate of magnesia with sulphate of iron and sulphuric acid affords an example of tonic and depurant agencies combined.

The cathartic powder given at bed time twice a week has an especial effect in clearing out the colon.

May 27th. After a treatment of three months, completely restored—went home.

The plastic treatment is considered essential to the result, but the protracted constitutional treatment is thought to be equally necessary.

This may be taken as an illustration of the impracticability of reducing the practice of medicine to the sharp divisions of specialties, adopted in many of the arts.

Hare-Lip, Stokes' Modification of Malgaigne's Plan*—Dr. Wm. Stokes, Jr., of Dublin, (see Dublin Quarterly Journal for August, 1870, and the American Practitioner for January,1871,) has proposed a modification of Malgaigne's plan. The original plan is distinguished by leaving the vermilion border of the lip uncut.

This permits the detached edges to hang down while the sutures are being applied for the closure of the fissure. After this closure, the projecting material is pared away to the requisite extent.

The object of this, is to obviate the occurrence of a notch at the place of joining of the prolabium. Dr. Stokes' modification consists in cutting only about two-thirds of the way through the thickness of the lip, except at the line of the vermilion border where the incision turns from the vertical

*Note to p. 62 of Plastics of 1867.

to the horizontal direction, and here the incision goes through. Above this horizontal line, the superficial two thirds, is dissected from the deep seated one-third on each side, and turned in toward its fellow.

Fig. 12.

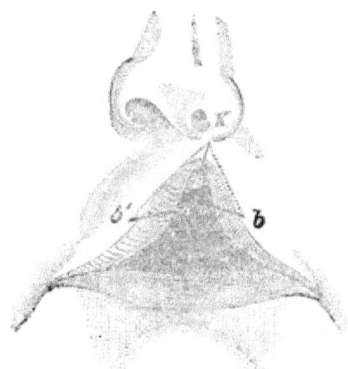

Fig. 13.

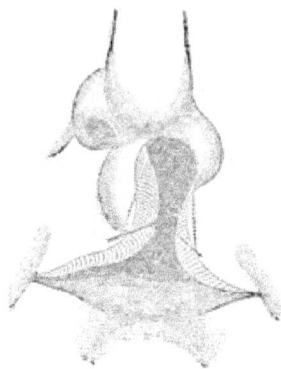

Stoke's modification of Malgaigne's operation for hare-lip.

B. The fissure.

K. The point of convergence of vertical incisions.

K *b* and K *b'*. Dotted lines indicating vertical incisions two thirds through.

a-b and *a'-b'*. Horizontal incisions entirely through.

The tissues embraced in the diagrams K *b a* and K *b' a* are turned in with their raw surfaces toward each other.

A modified conformation of the deformity.

When the sutures are applied, these inverted cut surfaces come in contact, greatly increasing the depth of the incised surfaces in apposition, and the consequent thickness of this portion of the lip.

The two projections of the vermilion border hang down, as in Malgaigne's operation and are treated in the same manner. A glance at the cuts will make this more plain. The dotted lines indicate vertical incisions passing two-thirds of the way through. The heavy black lines, the horizontal incisions passing completely through.

Destruction of the right cheek, the lower lip and half the upper lip. Reparation by transplanting integument from below, by the third method of the author's classification.*—At the time of making the report of 1867, the case of Mary Bowers, illustrated below, was incomplete. The final result; the production of a lip completely hiding the teeth and retaining the saliva, had not then been attained. The last of the

---

*Note to Plastics, 1867, p. 67.

series, of four illustrations, did not therefore appear in that report and they are all introduced here for the completeness of the exhibition. The destruction of the cheek followed an attack of *cancrum oris*. No mercury is said to have been taken in connection with the disease. Exfoliation of portions of the superior and the inferior maxilla, with the loss of several teeth, occurred in connexion with the gangrene of the soft parts.

Fig. 14. Destruction of the right cheek, the whole lower lip and half the upper lip.

Fig. 15. The condition three weeks after the first operation.

Fig. 16. The condition after the fourth operation.

The deformity is illustrated by Fig. 14, with the dotted line, showing the form and position of the flap elevated in the first operation.

Fig. 15, shows the condition three weeks after the first operation.

Fig. 17. The final result after the 6th operation.

Fig. 16, shows the condition after the fourth operation.

When it is proposed to close large cavities, it is often necessary to accomplish much of the detail in separate operations, permitting the parts to heal between times; very much as the modeler in clay permits his work to dry between his sittings.

Fig. 17, shows the final result of the sixth operation. A very good face is not only secured, but the patient's mouth is able to hold saliva, which is a rare attainment with a new-made lower lip. The lip is kept from falling down by the line of immobility about to be described. (See page xxx.)

Epithelioma of the lower lip and the left cheek.—The cuts,
(Figs. 18 and 19), show a plan which can often be made available in similar cases. The end to be accomplished is the
formation of natural lip-surface at the corner of the mouth,
presenting the regular amount of vermilion border without
inversion of the skin. It is especially important to avoid
this inversion in the male face, on account of the trouble occasioned by the beard.

Fig. 18, will afford a very good conception of an epithelioma
of one year's growth, in the cheek and lip of Mr. S. H., aged
40, of temperate habits and in other respects good health.
The shading is intended to indicate the thickness and hardness
of the morbid growth.

Various caustic applications had been made, by irregular
practitioners, which only seemed to accelerate the increase of
the induration.

A deep ulceration existed at the left corner of the mouth;
the outer left fourth of the lower lip was indurated without
increase of volume, and a considerable thickening,with induration, existed in the left cheek, the shape of which is indicated
in the cut (Fig. 18). It is convenient here to give the result
of microscopic examination.

The muscular tissue is everywhere present in the diseased
mass. There is no distinctly circumscribed deposit; but an in-
filtration of the normal tissues. Epithelial cells and scales appear
in all parts of the induration but no caudated cells. The part
had been the seat of occasional lancinating pains, but the examination shows the growth to be an epithelioma. The shaded
surface, included in the line *a b*, shows the extent of disease
involving nearly half the lower lip. The incision was made
to follow this line. The additional incisions, represented by
the dotted lines x x were made in the skin parallel and near
to the vermilion border, in order to give mobility to these portion of the lips.

Fig. 19, shows the appearance after the removal of the
tumor and the partial application of the dressing. The free
portion of the upper lip has been carried down and joined to
the free end of the fragment of the lower lip at *c*, by fine
cotton thread sutures smeared with carbolized wax and
cerate. This leaves a large opening into the mouth out side

of the newly constructed lip. In order to bring the cheek up to the lip, it is first freely dissected from the maxillary bone to give it mobility. Sutures are then introduced to bring the incised surface of the cheek forward, until its contact is secured with the lip. This process is shown in the cut by a plastic pin e, around which a thread d, is passed to be tightened. A considerable elevation is produced in the cheek above and below the pin as the opening is closed, and it is better to let them remain for they will diminish very much by distribution and equalization of tension, and after this process has ceased, any amount of integument may be removed which may be necessary to secure a smooth surface. If, on the other hand, there should be a failure to secure adhesion, any unnecessary destruction of the substance of the cheek would add to the deformity. No plasters were employed. The method of reinforcing the sutures with carbolized paper very much diminishes the necessity for plasters.

Two ligatures were applied to bleeding arteries in the cheek, cut short, and left to drop subsequently into the mouth. Notwithstanding this, the blood flowed too freely into the mouth after the completion of the dressing and the recovery of the patient from anesthesia.

The cheek compressor (Fig. 1, p. v.) was applied and worn for twelve hours. It completely arrested the bleeding. If it had not been for this instrument, great difficulty would have been experienced in controling the hemorrhage.

This wound united by the first intention wherever incised surfaces were in contact.

On the third day from the operation, nearly all the sutures and one of the two plastic pins were removed. In two days more the other pin e, passing from the newly made corner of the mouth across the cheek was removed. There was at no time any inflammatory redness along the margins of the incisions.

No plasters were at any time applied, but the surfaces were brushed several times a day with the antiseptic solution of carbolic acid, chloride of zinc and glycerine.

It ought to be mentioned in connexion with this history, that on the opposite angle of the mouth. Dieffenbach's opera-

tion (See Fig. 20,) was made in order to give the mouth
sufficient extent.

April 16th.—The patient returned for inspection, on account
of a slight ulceration in the vertical line in the lower lip,
where the upper lip had been united with the fragment of the
lower lip, marked c in the cut Fig. 19.

A fraction of a drop of carbolic acid was applied, and directed to be re-applied every day, and to await the result of this
before submitting again to the knife.

April 27th.—The operation was repeated in the same manner,
removing nearly as much tissue as before.   Union by adhesion
resulted, and the pins and sutures were all removed on the
third day.   An additional pin was introduced on the fourth
day to close a fistule at the lower end of the wound, through
which saliva dripped, and on the fifth day the patient went
home.

June 10th.   Up to this date, there is no return of the disease, and the deformity is nearly concealed by a moustache.

Fig. 18.   Fig. 19.

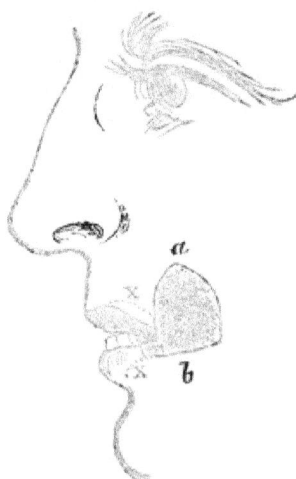

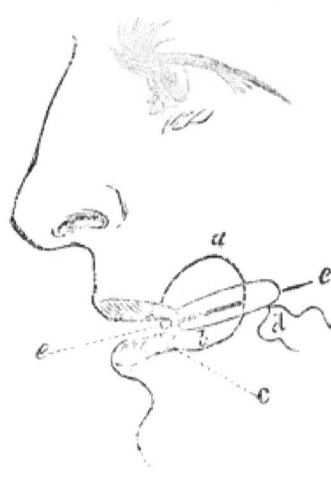

a b. The line of incision showing the extent of the morbid growth.
X X. Dotted lines showing the places in which incisions were made outside of
the vermilion borders of the lips.
e e. The principal pin employed in the dressing.
c. The line of joining of the fragment of the vermilion border of the lower lip, with
that of the upper lip brought from above.
d. The knotted portion of a thread passing under the head and the point of the pin
(e e.) indicating the manner in which the parts are to be approximated.

Deficiency of the Lower Lip..—Figures 20 and 21, are taken
from photographs of a case of deficiency of the lower lip, from
gangrene, said to be dependent upon the use of mercury at the

Fig. 20.

Deficiency of lower lip from mercurial gangrene.

age of two years. The operation was done April 9th, 1869, at the age of about twenty-five. The first method, second variety, was followed—the process is rendered readily intelligible by the dotted lines on the cut. The straight line across the chin is made deep to the bone, and the portion of the lip and cheeks above this line freely dissected away from the bone, to permit free mobility. The two tongue like portions are united in the median line, and the mouth is then extended on either side by the operation of Dieffenbach,* which consists of the removal of an area of skin and subjacent tissue, half or two-thirds of the thickness of the cheek, and the making of a horizontal incision through the mucous membrane, so as to preserve all its surface.

The dotted lines around the corners of the mouth illustrate the method. Very little deformity remained.

Fig. 21.

Appearance several weeks after restoration. From photographs.

Fig. 21, has been engraved from a photograph taken several weeks after the operation. The most important advice to give those who have not had experience, is to be sure and make a sufficiently free dissection of the cheeks from the bone, and to carry the incision close upon the periosteum, in order to avoid the arteries. This should be done

*See Plastics, 1867, p. 76.

with a strong, blunt-edged knife, made with reference to cutting upon bony surfaces.

A line of Immobility*—A New Method in the Restoration of eversion of the lower lip from cicatricial contraction.—A Flap also brought from the arm.—The importance of the method justifies a considerable detail in the report of the case.

Miss A. C., aged 23, was playing, when four years old, in front of a stove, from which her apron caught fire, and before it could be extinguished, the arms, neck, and lower portion of the face became badly burned.

The result of the consequent cicatricial contraction was an extreme depression of the lower lip, completely unfolding and everting it, and carrying it down to a level with the shoulders, as seen in the cut.

Fig. 22.

A. C. previous to treatment.

At the age of five years, or one year after the accident, and eighteen years ago, I operated upon the child, making extensive flaps, with their bases upward, restoring the lip to its proper position.

The next day after the operation, the symptoms of measles appeared, and the wound suppurated throughout.

The result was a failure, so far as the lip was concerned, but the depression of the lower eyelids was somewhat relieved. In this particular, the operation was a sufficient compensation to the patient for the suffering attendant upon the protracted suppuration.

The present condition is one of entire inability to close the

*The attention of the profession is invited to an element in plastic surgery, first publicly suggested by the writer in a monograph on orthopedic surgery, published by Lindsay & Blakiston, in 1866, enlarged from a report made in 1864 and repeated at greater length, with an illustrative wood cut, in a Report on Plastics, made to the Illinois State Medical Society, in 1867, and also reprinted as a monograph. The plan has also been briefly described in the Chicago Medical Examiner for January, 1870, and in the American Practitioner for April, 1871, p. 222.

mouth, except by an art the patient has acquired, of stopping
it up with the tongue.   The inferior maxillary bone is permanently bent by the downward pull of the cicatrix in front
and the upward lift of the masseter and temporal muscles be
hind. The lower incisors pointed straight forward; the tongue
had been educated to the capability of preventing the
loss of saliva.

Fig. 23.

A. C.   Six weeks after the commencement of treatment. The further depression
of the lip is prevented by the line of immobility under the chin.

Sept. 29th, 1870.   The patient having taken a cathartic the
day before and five grains of quinia the night before and
again in the morning, and being in a state of anesthesia from
chloroform; the dissection was commenced from below.   The
depressed lip, with a wide margin of sound integument from
the neck on either side, were separated from the subjacent
tissues and made to glide up onto the jaw. In this process, the
periosteum was scraped off from the base of the jaw, in order
that the integument restored to the chin, might not be drawn
down upon the neck, in the progress of granulation and
cicatrization.

It is this point, in connexion with this case, which constitutes the improvement in the surgical art.   The desideratum
has hitherto been to prevent the cicatricial traction upon the
lip in the process of healing.   By peeling off the periosteum
along the base of the jaw, contraction of the portion of the
cicatrix which lies upon the neck, is limited to this denuded

line upon the bone, and is fenced off from connexion with the lip.

Dec. 9th. After the lapse of nearly six weeks, the denuded surface upon the neck had nearly half healed over, and all suppuration upon the anterior surface of the chin had ceased. The entire absence of any sound integument below the vermilion border, permitted the lip to be unfolded again and drawn down upon the chin as far as the line of immobility at the base of the jaw, presenting the appearance shown in Fig. 23. The abundance of integument upon the sides of the face, gave promise of material for supplying the deficiency upon the chin.

Second operation.—Accordingly, a strip of integument three quarters of an inch wide and two inches long was dissected from the base of each cheek, the base being central. These were brought around (III. method 2nd variety) each making a turn of a half circle, one being placed above the other and completely covering the chin. These were retained by silver sutures and isinglass plasters so as not to produce tension or pressure.

Unfortunately, both the flaps sloughed, leaving the case worse than before. In the light of this experience, the application of the strips of plaster was a mistake, though there appeared to be no compression.

January 17th, 1871. After another interval of between five and six weeks, the health having been invigorated by riding out, cathartics quinia and iron; and after preparation by anesthesia, a flap with its base toward the shoulder was taken from the right arm and attached to the chin.

January, 23rd. Notwithstanding pretty good adhesion and all our care, the flap pulled off. The flap was further dissected from the arm and re-attached with the additional precaution of drilling a hole through the mental portion of the jaw and attaching four threads to a button under the bone; the threads coming forward through the orifice in the bone and spreading out, passing through the areolar substance under the flap, and being attached to buttons of lead covered with wax on the outer border of the flap. These buttons were ellipses in shape and perforated at each end, each orifice receiving one of the four threads. This held the arm to the chin with great security.

January 31st. The flap had adhered to the chin throughout

and preserved a good color, until these threads, acting as fasteners of the arm to the chin, were found to have cut their way so far into the substance of the flap, as to cut off the circulation and cause gangrene of about five-sixths of it. The remaining portion however, maintained a vigorous vitality. More flap was obtained from the arm by successive ligations until the separation from the arm became complete.

These ligations were made by, introducing from time to time, a plastic pin and passing and retaining a thread over it until the intermediate substance was cut through. Thus the integument attached to the chin became gradually accustomed to receive its blood through a narrower base.

February 17th. For 31 days the arm had been kept in apposition with the chin by the continuity of the flap, except an interval of a few hours between the tearing loose of the original flap, and its reposition. Upon the final separation, at this date, it was covered with cotton to prevent any accidental depression of temperature. After two days, this precaution was discontinued and the flap continued to maintain a healthy circulation through its new attachment to the chin.

Fig. 21.

A. C. A permanently restored lip with an abundant chin brought from the arm. From a photograph taken four months after the beginning of treatment.

For a long time, a portion of a small rubber tube was worn under the chin, and fastened over the head to prevent the coalescing of the granulations upon the neck with those under the base of the jaw. It was unexpectedly found that although the bone remained rough along the surface from which the periosteum had been scraped, yet the vigorous and redundant granulations would bridge over this barrier if their projecting surfaces were allowed to come in contact. During the latter portion of the treatment, a compress was also worn under the chin to push up the newly placed integument and prevent its forming a fold so as to overleap the line of immobility.

May 16th. There is still a small portion of unhealed surface upon the upper portion of the anterior surface of the neck. The implanted integument upon the chin has nearly lost its tendency to bag over in front of the neck and the orbicularis oris is constantly gaining, in its tendency to secure the permanent closure of the lips, without voluntary effort. June 29th. Nine months from the commencement of treatment. The lip maintains its position, independent of all artificial support. Only a very small surface upon the neck remains to be skinned over.

Altogether, this case affords a gratifying triumph over great difficulties; the horizonal teeth gradually assumed a perpendicular position, being lifted up by the pressure of the restored lip. The operation of Teale, building a new lip upon the old one, is the only other conceivable plan by which a voluntary closure of the mouth could be secured. The plan here pursued, ultilizes the natural lip with its proper muscle and affords to the neck the possibility of being restored to its natural narrowness.

Intestinal Plastics.—It is important that the conceptions of the plastic surgery of the intestines should be simple and free from the idea of the necessity for unusual or complicated appliances, because the emergencies which commonly call for surgical operations do not admit of delay. The first condition to be secured, after a wound of the intestine has been made by accident or design, is the placing in contact of surfaces which easily adhere. These are the serous surfaces of the portions of intestine to be held together. It might be thought that the cut surfaces should be placed in contact, but they are so thin that it is impossible, by any reasonable number of sutures, to prevent the escape of the intestinal contents through the interspaces of the stitches. The continuous suture is the only one which could be relied upon for the purpose of holding the incised surfaces together, and even with this suture, it would be very difficult to prevent inversion or eversion of the surfaces. The contact of the cut surfaces is the less necessary, as the serous surfaces adhere with quite as much facility, and with the advantage that there is no great care necessary in adapting the serous surfaces to each other. The interrupted suture is more in accordance with the fashion of surgery, but the continuous suture will here do as well, and there is this advantage in it, that as it is expected that the suture will ultimately pass off into the intestinal canal, the first part which ulcerates through to the mucous surface serves as a means of pulling upon the other portions, until finally the whole of it has ulcerated through. If however, it is intended that the sutures shall remain permanently in the tissues, they must be of silver, interrupted, and introduced with care not to perforate the mucous membrane. The ends must be cut off close to the knot or twist, in order that they may be easily covered. The best general direction doubtless is to make a continuous suture with a common sewing needle armed with a fine thread of silk, linen, or cotton which has been repeatedly drawn over the carbolized wax and then smeared with the carbolized cerate. (See page. vii.)

The capability of the suture for absorbing fluids is thus very much diminished and what is absorbed by the filament, is held back from going into the putrefactive change, until the antiseptic has been dissolved out, and by this time the suture has been well covered with plastic lymph, affording protection to the living tissue.

The introduction of carbolic acid into the material of the suture does much to diminish the necessity for the use of silver wire, and it is one of the improvements for which surgery owes very much to Lister.

The stationary condition of the wounded part. In the opinion of the writer, it is of great importance to attach the wounded portion of the intestines to the abdominal wall so that the wound in the one shall come opposite that in the other. The indication is to hold the injured portion of the intestine in one place, so that if the plastic exudation upon its surface proceeds to suppuration, there will be a greater likelyhood of its being walled in by the surrounding adhesion, and if there should be any leak of intestinal contents, the chances of their external escape may be greatly increased. This indication is generally disregarded by writers upon surgery, but it only needs to think of it, to appreciate the danger of permitting a wounded and inflamed portion of intestine to float unrestrained among the intestinal folds.

It needs no argument to show that surfaces will more readily adhere by the joint production of plastic lymph, than in the condition in which only one of them is the seat of adhesive effusion. This leads to the maxim, that, *in all cases in which a loop of intestine inflamed, or certain to become inflamed, is returned from an open wound into the abdomen, whether wounded or not, it should be restrained by a suture from floating away.* This applies to cases of strangulated hernia, in which a greatly inflamed loop is returned into the abdomen, after an external incision has been made through which a bistoury has been introduced, for the division of the stricture which prevented the reduction by manipulation.

Still more important is the observance of this rule, if the intestine has been accidentally wounded by the bistoury in the division of the stricture. A suture is supposed to be introduced so as to bring the sound surfaces together over the wound in the intestine, however small it may be, but the effusion of plastic lymph is essential to the complete reparation. If this fails, and the loop of intestine, instead of being retained at the abdominal wound, is floating loose in the abdomen, the danger of feculent effusion must be increased a thousand fold. It will generally be found convenient to attach this suture to the wound in the intestine and to bring it to the cutaneous surface, not through the abdominal wound, but through the neighboring tissues, in order to favor, as much as possible, the closure of the wound in the abdominal wall by adhesion. When there is no wound of intestine, as in most instances of operations for hernia, the suture can be attached to the mesentery or mesocolon and brought out through the abdominal wall in the same way. The theory is, that the suture will be gradually drawn to the exterior by moderate traction, cutting its way by absorbtion, so that the portion of the suture attached to the intestine or its retaining mesentery, is the last be disengaged through the skin.

For the purpose of such a suture, a silver wire possesses the obvious advantage of the absence of the possible generation of any putrefactive irritant.

Posture and Medicine.—The same necessity which indicates the use of a suture to retain the injured intestine in one place to favor adhesion, also indicates the necessity for perfect quietude in the patient's posture, in order to favor the ready adhesion of adjacent parts, and to avoid the breaking up of whatever union may have been already secured.

Opium is for the same reason universally indicated to arrest the vermicular movement of the intestinal loops among themselves. However constipated the patient may already be; no cathartic is on any account to be administered during four or five days while the adhesions are tender.

This discussion of principles is necessary, to settle the course of practice in those permanent displacements or deficiences of the intestines which require treatment corresponding with plastic operations in other parts of the body.

The most frequent of these, are the cases of intestinal fistule called artificial or adventitious anus.

The following case of intestinal plastics, detailing the use of a new instrument in place of Dupuytren's forceps for restoring the continuity of the canal in a case of intestinal fistule, is republished from the American Journal of the Medical Sciences for October, 1869.

J. C. C. æt 29, tall and slender, (colored clergyman,) entered my Infirmary Nov. 17th, 1868. He had become in May, 1867, the subject of inguinal hernia, with the symptoms of strangulation, which continued twelve days, when a distinguished surgeon of Iowa operated upon him, leaving the patient with an adventitious anus, and the loop of intestine adherent within the scrotum. Whether the incision of the intestine was made upon the supposition that there was gangrene, or whether it was accidental is not known. From subsequent examination, and the well-nourished condition of the patient, 'the seat of hernia seemed probably to be in the lower portion of the small intestine, in which the feces passed from right to left, and that the opening made by the surgeon was in the ascending portion of the loop. It follows that the feces all passed out of the abdomen into the scrotum, and, in returning into the abdomen, passed by the adventitious opening made by the operator. Much of the contents leaked out, especially when, in consequence of taking cathartics, or having a diarrhœa, they were unusually thin. The patient ordinarily wore a compress, with a complicated fastening, his own invention—which, however, was very uncertain in its security. An attempt was made by the original operator to close the orifice by a plastic operation in the following September, which failed, and he repeated the attempt in November, December and February without success. In none of these operations was any attempt made to dissect up, and explore the intestinal protrusion; but according to the testimony of the patient, the operation was only practiced upon the integument for the purpose of securing adhesion and closure of the cutaneous orifice. Bearing in mind the patient's account of the failure of preceding operations, it was resolved not to incur the risk of failure from the same cause.

My first examination of the case was intended to be thorough, but I failed to detect the entrance of the ilium into the external loop through the ring. It was supposed that the intestinal wall had so sloughed as to remove the partition wall, and that an operation which should sufficiently dissect the intestine from its adhesion in the canal, permitting it to be drawn into the abdomen, would lead to a closure of the orifice. The progress of the operation revealed the mistaken diagnosis, and led to a change in the plan of treatment.

Operation Nov. 19th, 1868.—The loop of intestine was first dissected out from the scrotum, and the portion of intestine protruding through the external ring was cut off. It then appeared that there were two intestinal openings into the abdomen, and the philosophy of the case was for the first time unequivocally cleared up. The accompanying cut (Fig. 25.) illustrates the anatomy of the case.

As Dupuytren's forceps for the gradual division of the septum were not at hand, a ligature was introduced through the septum, about an inch and a half beyond the level of the skin. Each end of the ligature was passed through a short double canula and made fast, and from day to day tightened up until it cut through. The fear of peritoneal inflammation prevented the insertion of the ligature to such a depth as certainly to restore the permanent continuity of the intestinal canal.

Velpeau attributes to Schmakhalden the original conception

of the plan of introducing a ligature for restoring the continuity of the intestinal canal, which was published in 1789. Dr. Physick's operation by ligature which he supposed to be original, performed in 1809, was followed by amendment, but failed of a cure, on account of the want of depth of the channel made by the thread. The edges were afterward made raw, and twisted sutures introduced; but, after apparent success, the new union was torn open by the pressure of the feces from within.*

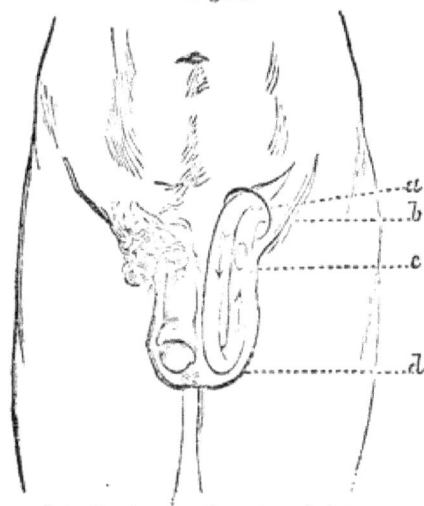

Fig. 25.

a. Intestine leaving the external ring.
b. Intestine entering the external ring.
c. Opening in the ascending portion of the loop of intestine.
d. Lower extremity of the intestinal loop in the scrotum. The arrows show the course of the intestinal fluids.

A good deal of constitutional fever followed this operation, the patient being delirious for several days. The rapidity of the pulse was kept down by veratrum viride at first, and the powers afterwards sustained by quinia, iron, and beef-tea. Some sloughing of the scrotum occurred, apparently in consequence of the arrest of pus in the pocket from which the intestinal loop had been dissected. The thread cut through in a few days, and the external wound contracted rapidly; but upon careful examination it was found that the septum came too near the surface to make it safe for the integument to close; lest a stricture and arrest of intestinal contents should be the result. Besides, it was found that there was still a bridge and an orifice beneath it from one portion of intestine to the other, from which it was supposed that there must have occurred adhesion in the septum or eperon behind the ligature. This supposition is the more probable as the surfaces, covered by granulations, pressed against each other, and would thus have

*Chelius' Surgery, with notes by South, vol. ii. p. 139.

the best opportunity to hook into each other, and thus effect a continuity of tissue.

In order to deepen the channel, the employment of the forceps of Dupuytren (enterotome) was now contemplated; but the history of the instrument, a few sketches of which I will quote, led to the adoption of a modification of his plan. (Fig 26.) Dupuytren commenced with the employment of a ligature, but, dissatisfied with this on account of its danger and uncertainty, he devised a forceps, to be gradually tightened by a screw, in order to cut through from one portion of intestine to the other. Out of 41 cases—21 by Dupuytren, and twenty by other surgeons—3 died.

"The adhesive inflammation," says Velpeau, "does not always take place at the periphery of the enterotome, even after it has been applied in the most judicious manner. In some cases it is almost impossible not to include between the branches of the instrument a portion of some important organ at the same time with the abnormal septum. Finally, in many cases artifical anus and stercoral fistules will persist to an indefinite period of time, in spite of the destruction or absence of every kind of eperon."

Velpeau relates a case in which death occurred at the end of the eighth day after the introduction of the forceps of Dupuytren, though the patient went on very well during the first four days. On examination it was found that gangrene had been produced by the pressure of the forceps, and the failure of adhesion had permitted the contents of the intestines to be poured into the peritoneal cavity.

The same surgeon relates another case, in which an applicant for an operation died of erysipelas without having had an operation; and on examination, it was found that one intestine was twisted around the other, so that the forceps would have endangered the effusion of feces into the peritoneal cavity, as actually happened in the other case from want of adhesion between the peritoneal surfaces.

"Frequently, after the destruction of the eperon by the method of Dupuytren, the artificial anus persists under the form of a fistule, which cannot be closed by any means applied."*

The contemplation of the danger of failure of adhesion after the introduction of the forceps of Dupuytren, led to the invention of an apparatus intended to avoid gangrene. For this purpose it is necessary to avoid a tight squeeze upon any of the tissues. A hook or tongue is made to invaginate the

*Velpeau's Surgery, by Mott, vol. iii. pp. 632, 634.

intestinal coats through a ring, thus bringing their peritoneal surfaces into close contact, but without such force as to interfere with the circulation. The perforation takes place by a gradual thinning over the point of the instrument, so that the orifice is at first small, and is surrounded by a large extent of serous surfaces in close contact. This differs entirely, in the principle of its action, from the instrument of Delpech, which cuts out a disk by the gangrene occasioned by the pressure of two rings together, involving more risk than by the forceps of Dupuytren, which only cuts a fissure.

Fig. 26.

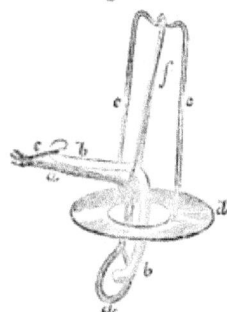

The action of the instrument will be better understood from the cut (Fig. 26.)

The apparatus consists of—

a a. A loop or ring to be introduced into one portion of intestine through the orifice.

b b. A perforating hook for the purpose of making a communication between two adjoining intestinal tubes.

The engraver has made the hook too short—It should project through the ring.

The loop or ring having been introduced through one intestinal orifice, and the hook through the other, the hook or male part of the apparatus pushes a portion of the double intestinal wall through the ring or female portion, and slowly perforates the intestine by ulceration without gangrene. Two peritoneal and two mucous surfaces are to be perforated by the point or hook invaginating them within the circumference of the ring. As there are no sharp corners or points, the process is sufficiently slow to permit adhesion of the peritoneal surfaces. The opposite end of each horizontal portion of the apparatus has a hook to hold an elastic cord to aid in the pressure of the hook through the loop.

c. The elastic cord just mentioned.

d. A shield made of tin, to serve as the base of a lifting process to be instituted as soon as the hook *b* has fully engaged in the loop *a*.

e e. A derric for the lifting process.

f. An elastic cord attached to the combined arrangement *a a*, *b b*, and tied over the top of the derric *e e* as the lifting power.

In two weeks a passage seemed to have been made from one

tube to the other, through which some of the intestinal contents passed. It become necessary to apply a lifting force to the hook and ring, in order to force them to divide the bridge lying between them and the surface. For this purpose a derric (Fig. 26, d, e e, f,) was constructed with a base of tin having an orifice in the centre, with a loop of wire about three inches high. From the top of this loop an elastic cord f was extended to the wire apparatus, constituting the hook and eye, by which a deep orifice had been made from one portion of intestine to the other. When the hook had come very near to the surface, a ligature was passed beneath the bridge, and having been passed through the tubes of a ligating canula, it cut through in a very short time.

After the complete and ample restoration of the continuity of the alimentary canal, the external orifice diminished rapidly, but at length it came to a stand-still. Finally, to close this, a plastic operation was performed Feb. 23d, 1869. This consisted in a free dissection of integument around the orifice, and then the bringing down of a flap of integument from above by the first variety of the third method of my classification[*]— that is, by a jumping process. The flap was carefully adapted to its new position, and retained by sutures of iron wire. Moderate compression was employed to prevent its separation by the pressure of the feces beneath. The surface from which the flaps had been taken, was left to granulate and cicatrize. Adhesion was effected in every part, and the final cure was thus secured after a period of treatment of three months' duration. The patient was advised always to wear a truss to protect the part from the danger of a hernial protrusion from the pressure of the intestines upon the enlarged ring. The removal of such a horrid disability could not fail to secure the warmest gratitude of the patient.

Velpeau, in his work on Surgery, refers to a case of autoplasty by M. Collier, in which a cure was effected by detaching a portion of skin from the neighborhood, and attaching it over the orifice by means of pins. The implantation of natural integument over the orifice has the advantage of providing material which is not likely to be torn apart by a considerable degree of distension produced by lifting and other efforts At this date, May 1871. The patient continues well and is traveling as an itinerant preacher.

Fission and Extroversion of the Bladder and Epispadias.* —Eight cases have been treated by plastic operations by John Wood, F., R., C., S., &c. (Lancet, Feb. 20th, 1869, and half yearly abstract No. 49, for July, 1869.)

After trying various methods of sliding, he came to the adoption of the plan of inversion by which skin is made to take the place of mucous membrane.

Surgeon Wood attributes to Pancoast the credit of originating the idea of inverting the skin to make a substitute for mucous membrane in an artificially constructed bladder.

He thinks that the action of the urine upon the integument is to diminish the growth of the hairs, but in one case the patient was obliged to pull them out with forceps as often as phosphatic concretions formed upon them.

A weak solution of nitric acid was found to be of great service in keeping the surfaces clean.

In the same direction, it is stated by Prof. Simon† that acid urine is harmless when brought into contact with the tissues, while alkaline urine is highly destructive to surfaces not protected by an epithelial covering.

Pure acid urine has been injected under the skin of rabbits without any apparent bad effect. Operation wounds moistened by fresh acid urine healed by primary adhesion.

When ammoniacal urine was injected, even if it had been filtered, abscesses resulted.

"In plastic operations on the urinary or sexual organs, it is unnecessary to leave a catheter in the bladder so long as the urine is acid,‡ while such operations should not be performed, if possible, while the urine is alkaline."

It follows from this, that a weak solution of sulphuric or nitric acid should be drank in connexion with these operations for several days. A weak solution, merely tasting sour, may also be employed as a lotion, used as an injection, or applied to accessible parts by means of sponges.

---

*Note to Plastics 1867, page 89.

†Chicago Medical Examiner May 1871 p. 310, from Deutsche Klinik.

‡The use of a catheter, after the insertion of sutures into the bladder, is to save the seam from mechanical strain, and it ought therefore to be worn or frequently introduced for several days.

ORTHOPEDICS.

There is a growing appreciation of the importance of mechanical appliances, in the treatment of those diseases which are liable to lead to the distortion and disuse of parts. The notes and illustrations which follow are in further elucidation of the subjects embraced in the former report.*

Cicatricial Contraction.—It may not be improper to call the attention of the profession to the importance of mechanical restraint in the treatment of burns upon the limbs. While the cicatrix is in process of formation, there is a pull upon the surrounding skin nearly equal in every direction. It is during this period that mechanical restraint should be employed to prevent distortion. It is abundantly proved by experience that in most cases the tendency to distortion of the limbs arising from the demand of the contracting cicatrix for contributions of sound skin from every direction, can be controlled until this demand is lost in the equilibrium of tension.

The mistake of many practitioners, however, is in the too early discontinuance of the restraint.

It is not enough to see a complete cicatrization of the ulcer. The cicatricial contraction goes on for a long period after this, and even increases in power as the vessels which had carried the materials of repair become diminished in size.

During the period therefore, in which a cicatrix is passing from red to pale in its color, contraction is in progress, and mechanical restraint should be imposed. In the practice of this restraint, the splint or apparatus should be daily removed, and the limb subjected to all its natural movements in order to preserve the proper functions of its joints.

Great disappointment has resulted from the too early discontinuance of mechanical restraint, and this is the reason for the emphatic manner of stating the indication.

Vertical Curvature of the Spine.†—The abandonment of the plan of lifting in caries of the spine occurring below the cervical vertebræ, was a great advance in both the theoretic and the practical view of the subject.

---

*Report on Orthopedics presented to the Illinois State Medical Society in 1864 and enlarged and reprinted in 1866 by Lindsay and Blakiston. The references will be made to the pages of the book bearing the imprint 1866.

†Note to Orthopedics p. 126 of 1866.

The indication is, to apply a support like that which a splint affords in a transverse fracture of a bone. The articulating surfaces of the oblique processes, which are rarely or never diseased, afford an effectual stop to the direct shortening of the vertebral column, so that a splint which prevents the curvature, necessarily prevents the shortening. The splint is applied along the spine with fastenings around the pelvis below, and the shoulders above, and as long as the splint is firmly retained; the pressure in the erect posture, of the parts above upon the bodies of the vertebræ below, and the consequent bending of the diseased spine, are as effectually prevented as though there were a direct extension.

The pioneer in this improvement was Dr. Henry G. Davis of New York, and the accompanying cut (Fig. 27) illustrates

Fig. 27.

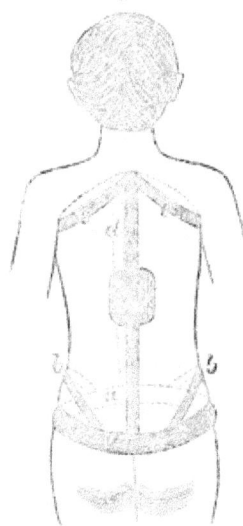

Early plan of Dr. H. G. Davis for vertical curvature.
a. Pelvic metallic band.
b. Oblique band over the iliac crests.
c. Metallic vertical splint supporting a pad f to apply against the projecting portion of the spine.
d. Dotted line showing the position of the anterior support.
e e. Axillary bands attached to the upper end of c behind and of d in front.
f. The pad which is applied to the bulging vertebræ.

the history of his earlier efforts. The idea of lifting had not been entirely relinquished at the time of the adoption of this plan:—the oblique bands b b passing over the iliac crests, being lifting supports against the pressure of the arms upon the axillary pieces e e.

Dr. Davis gave up the idea of lifting, and the profession has become nearly unanimous in its disuse. Notwithstanding the absence of the necessity for this lifting and the difficulty of securing it, nearly every instrument maker's shopwindow exhibits apparatus with crutchheads. The continuous pressure against the axilla is very uncomfortable and it will be found in practice very difficult to enforce the wearing of any appliance which does not afford more comfort than discomfort. I have known many cases in which some crutch-head apparatus was applied; but I have never known one worn long, without such a rebellion on the part of the patient, as resulted in its ultimate discontinuance. On the other hand, it is the exception to the rule, for a patient to rebel against a

lvi   ORTHOPEDICS.

carefully fitted apparatus which leaves the arms free for all
their movements. The plan illustrated in figure 59, page 146
of Orthopedics of 1866, is the most easy for extemporaneous
construction.

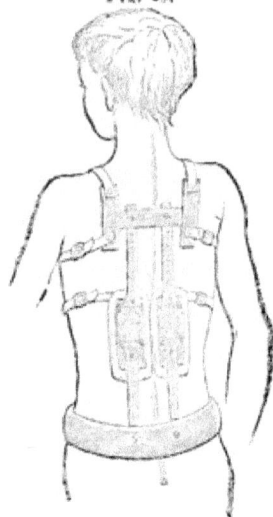

Fig. 28. Dr. C. F. Taylor's
apparatus with a joint against
the pads and a set screw to
increase or diminish the pressure upon the projecting portion of the spine.

Fig. 29. A front view of the
same. The shoulders are seen
to be held back, but not lifted
up.

Very beautiful and comfortable plans have been devised by
Dr. C. F. Taylor, of New York. The accompanying
illustrations Figs. 28 and 29 will afford a very good conception
of them.

A plan of adapting sole-leather is favorably mentioned by
Dr. J. S. Sherman of Chicago, and two wood-cuts, illustrative
of the application, were published in the Chicago Medical
Examiner for Oct., 1869.

If very much molding is required, it is necessary first
to make a plaster cast, and shape the wet leather upon it. If
however, only slight change in the shape of the leather is
necessary, it may be done directly upon the body of the
patient. When the leather gets dry, it affords very considerable resistance while it is at the same time slightly yielding.

Cervical Vertebræ.—Caries and curvature of the cervical
vertebræ require more complicated apparatus, and that which

necessarily restrains the natural movements. The plan of Sheldrake, of nearly a hundred years ago, has been variously modified. The principle consists in swinging the head upon a crane, which is attached to the trunk behind, and (curving forward over the head,) receives the swing in which the head is supported. Kolbe's apparatus for this purpose is shown in Figs. 30, and 31, upon Sheldrake's plan.

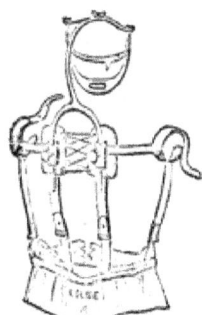

Fig. 30. Kolbe's apparatus for cervical caries and curvature. The principle of the swing is employed; the spinal column being the axis of rotation.

Fig. 31. Kolbe's apparatus applied.

Fig. 32. Dr. C. F. Taylor's apparatus for caries and curvature of the cervical vertebræ.

Dr. C. F. Taylor has devised an apparatus for cervical curvature in which, instead of the principle of the swing, that of the swivel is adopted. This is very well shown in Fig. 32.

The natural movements of the head must be more interfered with than with the use of the swing, but in wry-neck without caries, the greatest efficiency may require this.

A simpler arrangement may be employed for wry neck which consists of two ellipses, one for the head and the other for the chest. (See Fig. B.)

These are made of flat truss steel, or in the absence of this, of hoop iron, properly padded. The one for the chest must be made to open in taking off and putting on, and the one for the head must be attached to a cap to prevent its descending too low, and provided with a chin strap, to keep it from rising. Spiral fastenings of elastic rubber webbing ascending and descending between the two ellipses, suffice for turning the chin to the right or to the left.

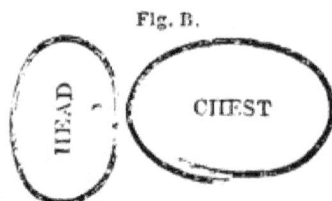

Fig. B.

HEAD

CHEST

Lateral Curvature.—Nothing has been introduced upon the subject of lateral curvature since the last report[*] to improve or simplify the theory or the practice.

Constraint against movement in the vicious direction, with freedom to go in the right, is true as a maxim, here as in other training.

Extension in joint inflammations. Cases are constantly coming under notice, presenting deformities which are extremely difficult to overcome and which would have been entirely prevented by moderate extension applied during the period of acute inflammation.

It is obvious therefore, that in the treatment of inflamed joints by extension, the prevention of those deformities which it is afterward so difficult to overcome, is equal in importance to the arresting of the disease itself. Fortunately both objects are favored by the same means.

A very moderate force counteracts the muscular contraction, at the same time that the joint surfaces are slightly separated, so that direct adhesion between the surfaces; or the impediments to mobility which arise from true or false ligamentous fastenings, do not render a limb useless as is the case when extreme flexion is permitted to take place.

By the general adoption of this treatment, weeks or months instead of years suffice for the return to health, while slight suffering only is endured, in place of the protracted agony which attends the disease when treated in the old way without extension.

Extemporized Apparatus for Hip Disease.[†]—In cases of hip disease characterized by intense irritation and sensitive-

---

[*]Orthopedics. 1866.
[†]Note to the article on hip disease in Orthopedics 1866, page 79.

ness to jars and movements, no kind of portable splint is to be relied upon. Nothing but the horizontal posture with weight and pulley, will effectually relieve the hip joint from the irritation of pressure and friction.

In the incipient period of the disease however, it is impracticable to hold the patient to the horizontal posture, because neither patient nor parent can be made to see the necessity for it. The same is also true of the period of convalescence. It is in these cases not easy to estimate the importance of having a plan for the construction of apparatus which any practitioner who has access to a blacksmith shop can easily execute.

The accompanying cut (Fig. 33) illustrates such an apparatus. It is composed of a shaft of iron, either solid or made of gas pipe, with a vibrating figure of 8 at the top, for the attachment of the perineal band, and terminated at the lower end by a double curved hook to hold the elastic strap by which the extension is secured upon the leg. This hook is reversed upon the shaft and is made concave, not only on its lower aspect, but also upon its anterior; so that the hook fits around the lower part of the leg. This curvature requires the apparatus to be made right, or left, according to the side which is the seat of disease.

A concave piece of tin, with a hook upon its convex surface near the top, is attached to the leg by means of adhesive plaster, cut so as to be about two inches wider than the tin and two-thirds of its length longer. The adhesive surface of the plaster is applied to the leg, and the smooth surface to the concave surface of the tin: the lower end of the plaster being reversed upon the convex surface and folded in upon the sides, so as to make a pocket for the tin. Adhesive strips or a roller, passed repeatedly round the leg and over the tin complete the attachment.

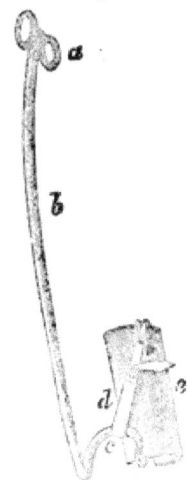

Fig. 33. This is an apparatus for extension in hip disease.
a. A figure of 8 attached to the shaft by a loose joint permiting vibration
b. A hollow shaft.
c. A hook which is reversed upon the shaft and made concave in the anterior direction to apply to the tin splint e.
d. An elastic strap with a buckle, attached to the hook of the shaft below and, to a small hook on the tin above.
e. A tin splint bent in the form of a half cylinder for adaptation to the posterior aspect of the leg. The convex side is presented. When applied to the leg it is reversed and brought forward of the hook at the lower end of the shaft.
The perineal band to be attached to the figure of 8 is not shown.

An elastic strap or cord is hung above upon the hook attached to the upper portion of this tin, and is fastened below to
the hook on the end of the shaft.

The strength and tension of this strap are the measure of
the degree of force in the extension. This is either increased
or diminished by the buckle or tie with which the strap or
cord is provided.

The shaft should be of such a length that the hook may press
upon the convex surface of the tin and not upon the leg lower
down.

It will be at once seen that it is not necessary to have a
very nice adaptation of length. The excellencies of this extending apparatus, are its easy and cheap construction, its easy
adjustment, its lightness and comfort, and its universal appli-
cability.

The tin attachment for the leg is a modification of that
devised by Barwell for the treatment of talipes.

In the cut, it is represented with its convex surface against
the convex aspect of the hook which terminates the shaft, in
order that it may be better seen.

Division of Tendons.—In all those cases in which deformity results in connexion with the destructive inflammation of
joints, there is a mass of cicatricial deposit which cannot
be reached by the knife without danger of exciting a deep
seated inflammation which may frustrate all the efforts for
improvement and lead only to disappointment.

The most ready illustration of this subject, is the case of a
knee joint which has freely suppurated, and which, after being
allowed to become flexed, has healed with extensive cicatricial adhesions. In such a case, the muscles flexing the leg
have contracted, but they present such a minor portion of the
resistance, that their division goes but a very little way in
overcoming the flexure. Force must be applied, and unless
the operator is satisfied to consume a very long period, it must
be so applied as to tear the cicatricial tissues. Force enough
for this purpose will also stretch the muscles as far as it is.

Besides the danger of deep incisions, every day's experience lessens the estimation of the value of the division of those
tendons which are superficial. Tenotomy is such a small element among the means of accomplishing the object, and the

wounds necessarily made are so apt to suppurate and become troublesome complications, that the increased facility of straightening, hardly pays for the trouble which the wounds occasion.

Direction of continuous force.—It is not always sufficiently borne in mind that the direction of continuous extension should be such as not to produce pressure of one joint surface upon another.

This is illustrated in Fig. 11. page 68 of Orthopedics.—1866. The illustration shows the tibia flexed and drawn back upon the femur so as to make an unnatural projection of the knee. The same is also well shown in a cut in Erichssen's Surgery.[*]

Fig. 34.

In this case the continuous extending force must be applied near the upper end of the tibia to avoid the inflammation which may be set up in those joint surfaces which are in contact, and the pain arising from the pressure communicated through the cartilage to the bone beneath whose sensibility becomes soon exalted by the attendant congestion.

Ankle joint.—Inflammation of this joint is very frequent, and there is a very clear indication for extension, but the practical difficulty is to make an attachment to the foot without injurious or uncomfortable pressure upon the diseased or injured joint. Dr. Sayre, of N. Y., has devoted much attention to the contrivance of expedients to meet this indication. The most complete apparatus for this purpose to be employed where motion is at the same time admissible, is that of Dr. Boisnot, of Philadelphia.

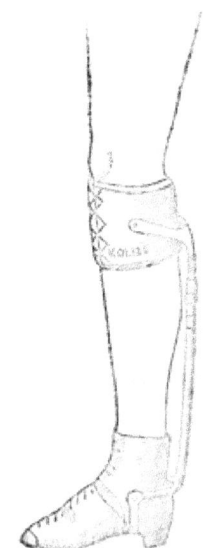

Fig. 33. Dr. Boisnot's apparatus for extension at the ankle joint.

The extending force acts through a gaiter.

The counter extension is secured by a broad band laced below the knee.

The extension is increased or diminished by a screw in the shaft which is behind the leg.

For an illustration of this I am indebted to D. W. Kolbe, manufacturer of surgical instrument, No. 15 South 9th St. Philadelphia. (See Fig. 34.)

It will be obvious upon inspection that the gaiter shown in the cut can be replaced by a strapping of adhesive plaster over

[*]American Ed. 1859 Fig. 252 p. 674.

the foot, and under an unyielding artificial sole applied to the sole of the foot. ·The counter extending band can be replaced by the tin splint figured in connexion with the apparatus for hip joint extension, (see Fig. 33.) only that in this case, it must be turned the other end up, so that the hook shall be at the lower end of it. With these modifications, the apparatus can be simplified so as to be constructed by any blacksmith.

Extemporized Apparatus for Talipes Equinus.*—It will be borne in mind that in most cases there is a doubling of the foot at the waist, or at the joining of the calcaneum with the cuboid bone on the outside of the foot, and of the astragalus with the scaphoid, and by this medium with the cuneiform bones on the upper and inside. This arching of the instep becomes firm by means of the shortening of the ligaments on the plantar surface, by which these bones are held together.

They are entirely beyond the reach of any cutting instrument, and they are too strong to be torn by any sudden force which it is proper to apply.

The indication is to apply force to the plantar surface of the metatarsal bones. The tibia is the fulcrum of this lever and the shortened tendo achillis is the resistance.

As one of the objects to be accomplished is the straightening of the crooked lever (the foot being the lever,) it is important that there should be a pretty firm resistance at the heel; any diminution of the force with which the tendo Achillis resists the pressure upon the metatarsus, by so far diminishes the only means by which the surgeon accomplishes the straightening of the foot. The division of the tendo Achillis is therefore worse than useless, unless there is an absence of the usual curvature at the waist of the foot. The high instep which is often seen after the treatment of Talipes equinus by division of this tendon is thus accounted for. Hereafter, with the general abandonment of this treatment by tanotomy, a more natural shape of the instep will be secured by the time the heel is brought down. We are now ready to appreciate the principle which should control the construction of apparatus. It is simply that of a lever.

---

*Note to Orthopedics 1866, p. 189.

Fig. 34 represents one of the forms which the lever may be made to assume. While the apparatus is attached to the sole of a shoe so as to bring the pressure under the metatarsal bones, a strap passes over the waist of the foot which throws the upper end of the apparatus forward of the leg. This upper end is then drawn back by means of a strap passing behind the leg.

A very powerful traction upon the tendo Achillis is thus obtained without interfering with the locomotion of the patient. Indeed, the motion incident to walking is an advantage as the ligaments and tendons yield more readily to a tension which is constantly varying, than to a steady pull. The reason of this is, that a much greater tension can be endured for the moment followed by partial or complete rest, than when it is continuous.

The cut (Fig. 35) represents an easy method of meeting the mechanical indication which has just been considered. It is the skeleton only of an apparatus. Neither the leather fastenings nor the enclosed foot and leg are shown. The imagination can see them.

Talipes Plantaris.*—The unusual case of Talipes plantaris illustrated in Figs. 36 and 37 occurred as the result of a scald. A child 3 years old accidentally plunged her right foot into a kettle of boiling water. The shoe protected the foot, but the leg was injured all the more from the hot water retained in the stocking.

As cicatrization went on, the contraction elevated the outside, producing a valgus, and the rising of the whole metatarsus, causing an apparent projection of the anterior portion of the calcaneum and of the cuboid making a plantaris. The cicatricial folds, which are very well shown in the cuts, acting upon the heel behind, and the metatarsus in front, hold the

Fig. 35.

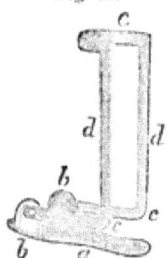

a. The sole of a shoe, the upper part is to be imagined.

b b. A flat and thin plate of iron attached to the sole. The turned up ends of this plate are perforated for a joint.

c c. The angle of a metallic strap, the horizontal part of which is parallel with the sole, and the vertical part with the leg.

d d. The vertical portion of the strap. This is a lever with fulcrum at c. The resistance at b, while the power is applied around the leg.

e. The metallic bow which connects the two levers at the top. A leather strap passes across the instep and has its attachment at c c.

*Note to the Article on Talipes in Orthopedics, 1866 page 151.

foot with such force as to neutralize muscular contraction, nearly arresting the natural movements.

A large ulcer seen upon the inside just above the ankle and extending half around the contracted circumference was the occasion of great suffering and trouble, and led to the determination to have the limb amputated. This was done in 1869 at the age of 20 years.

Fig. 36.   Fig. 37.

Fig. 36. Talipes plantaris occasioned by cicatricial contraction from a scald of the lower portion of the leg. An ulcer above the ankle covers half the circumference. Amputation at the age of 20, or 17 years after the accident.

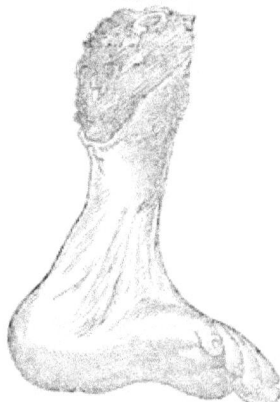

Fig. 37. The same viewed from the outside. The outer side of the foot is elevated disproportionately to the inner, giving the aspect of valgus to the deformity.

This is an instructive illustration of the power of cicatricial contraction. It is impossible that any surgical treatment could have obviated the resulting deformity. Again the compression of the arteries passing this cicatricial isthmus was such as to diminish the nutrition of the foot, and cause it to be smaller than the opposite.

A substitute for the hand.—It has been a desideratum to provide a substitute for the hand which will enable a person to hold a pen or pencil, and with the same instrument to grasp the handle of a hoe or shovel. In 1864 I explained my plan to Mr. Stohlmann, of the firm of Tiemann & Co., of New York. Mr. Stohlmann entered with enthusiasm into the idea and produced an instrument which combined the two points of usefulness, viz, the capability of grasping small objects like pins and pencils, and objects of considerable size, like the handle of an axe or hammer, a whip, &c.

Fig. 38.

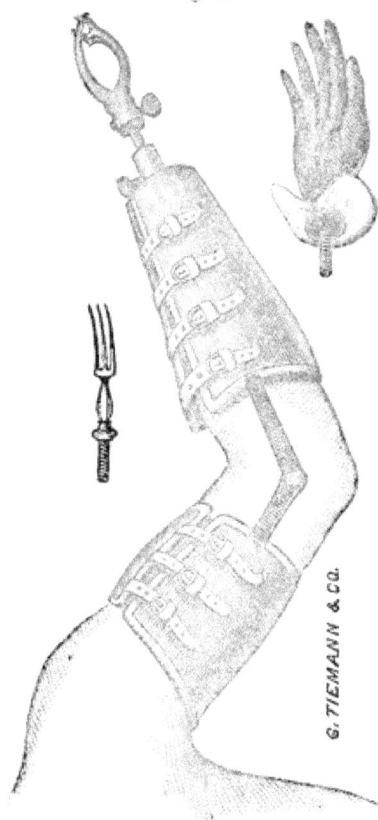

A conception of the instrument will readily be obtained by a glance at the cut (Fig. 38.) The apparatus is fastened to the investment of the limb by means of a screw, and it may readily be replaced by a fork, by the aid of the other hand, if it is not preferred to hold the fork in the grasp of the instrument, or a gloved hand can be screwed on in place of the useful instrument.

The elliptical opening is for holding a whip or hammer, and its size is regulated by a screw so as to fit objects of different sizes. Small objects are held in the fork at the end of the large opening, and by screwing down more or less the lever which constitutes one side of the opening, the grasp can be accommodated to the size of a pencil or a needle.

The instrument is represented in the cut as not quite closed for very small objects.

The screw, which constitutes the shank of the instrument, is so arranged that it can be made stationary, a necessary condition in holding a whip or a hammer, or left to rotate, as would be necessary in holding the handle of a plough.

It will be seen that the instrument has a great variety of adaptation.

There is an elastic strap which is attached to the chest to hold the investment of the arm, and to prevent the whole apparatus from sliding off in pulling or in carrying a weight.

The instrument is very far from being an adequate substitute for the natural hand, but it is believed that it combines the utility of a grasp for large objects with the capability of holding small things, better than any instrument which has yet been devised.

Conclusion.—This report may have failed to include many real improvements, the authors of which may feel disappointed not to find noticed in these pages.

In apology for this, it needs only to be suggested, that the report has been put together by arranging the half-hour jottings occurring in the intervals of practice, and any serious omissions would have been avoided, had attention been directed to them.

The chief aim has been to add to the resources of the Art—to make the path easier to tread by those who are to follow.

Note to page xliii.

While this page waits for the press the successful construction of a urinary bladder in a girl 6 years old, by Dr. John Ashhurst, Jr., comes to hand in the July number of the American Journal of the Medical Sciences, p. 70. The same number p. 154 contains also the report of two successful operations by Dr. F. F. Maury upon boys. In Dr. Ashhurst's operation the internal flap is everted from above, so that the cuticular surface becomes internal and the raw surface of this flap is covered by a flap brought from either side; the flaps sliding in curved lines.

In Dr. Maury's operation, the internal flap is brought from the perineum; a hole made in the base of the flap permitting the rudimentary penis to pass through as the flap is turned up, and the raw surface of this is covered by a flap brought directly down from above; the flap sliding in a direct line. The sutures of silver, called "The tongue and groove suture of Pancoast," employed in these operations is the same as that illustrated in Figs 10-11 page xix of this report.

The following table of 20 operations is copied from Dr. Ashhurst's article:

" *Table of Twenty Cases of Plastic Operation for Exstrophy of the Bladder.*

Operator.	Reference.	Whole No. of cases.	Suc-cessful.	Fail-ures.	Died.
Richard,	Wood, Med.-Chir. Trans., vol. lii. p. 96,	1	...	...	1
Pancoast,	N. A. Med.-Chir. Review, July, 1859,	1	...	...	1*
Ayres,	Am. Med. Gazette, Feb. 1853,	1	1	...	....
Holmes,	Surg. Treatment of Children's Diseases, 2d ed. p. 153,	5	2	2	....
Wood,	Med.-Chir. Trans., vol. lii. p. 85,	8	9	1	1*
Maury,	Am. Journ. of Med. Sci., July, 1871,	2	2	....	....
Baker	Med.-Chir. Trans., vol. lii. p. 187,	1	1	....	....
Ashhurst,	Am. Jour. Med. Sci., July, 1871, p. 70,	1	1	....	....
	Aggregate,	20	14	3	3

* Died from causes unconnected with the operation."

www.ingramcontent.com/pod-product-compliance
Lightning Source LLC
Chambersburg PA
CBHW022012190326
41519CB00010B/1490